I0570766

# Body-System Adventures:

## The Hidden Worlds Within Series

### Tom, Erika, and Billy Explore the Respiratory System

**L.D. Sledge**

**and**

**Dr. Shannon W. McPherson**

*Body-System Adventures: The Hidden Worlds Within Series*

**Tom, Erika, and Billy Explore the Respiratory System**

L.D. Sledge, and Dr. Shannon W. McPherson

Copyright ©2024 LD Sledge

Published by Dad & Dau LLC

Edited by Ita de Groot

First Printing

ISBN: 979-8-9897371-7-8

Library of Congress Control Number: 2024918066

sledgewriter@gmail.com

http://www.thebestghostwriter.com

doc@docmcpherson.com

http://www.docmcpherson.com

# Note from the author:

"Never pass a word you don't fully understand. If there is any doubt or consideration, always look up the definition in a good dictionary. Many times, I had the wrong definition and couldn't fully understand its use or concept until I "cleared" its definition. Each word may have different meanings, and you must pick the right one that fits the context of the sentence it is in. Clear its other uses as well as the right one. Use it in sentences and learn its derivation. Make it your own. This is more than good advice—it's the death of the reader to avoid it."

L.D. Sledge

# Table of Contents

# Prologue

After their incredible journey through the digestive system, Tom, Erika, and Billy were ready for their next mission—an urgent and personal one. Dr. Janus, their brilliant but ailing chemistry teacher, was suffering from a relentless cough that baffled doctors and left him in despair.

With the help of Merlin, the mysterious time traveler, the trio would once again be miniaturized, this time to explore Dr. Janus' respiratory system. Their goal: uncover the cause of his illness and find a cure.

As they journeyed through the nasal passages and trachea, they marveled at the intricate design of the lungs and the life-sustaining process of breathing. But deep within the alveoli, they encountered a sinister entity—an insidious disease impervious to conventional treatments.

Despite the danger, Tom, Erika, and Billy pressed on, navigating the lungs' labyrinthine of passages and witnessing the vital exchange of oxygen and carbon dioxide. Amid the battle against the disease, they discovered a surprising natural weapon that could save Dr. Janus, offering hope where science had failed.

Join them on this perilous journey into the respiratory system, where every breath counts and the fight to save a life reveals the wonders of the human body and the resilience of nature.

# CHAPTER ONE

## The Adventure Begins

Tom, Erika, and Billy stood on a little hill in Willowdale Park holding paper airplanes while looking down the clearing between the tall willow trees that graced the sides of the rolling hill they were standing upon. It was a beautiful day with billowy clouds immersed within a blue sky.

Tom, the tallest, said, "Mine will beat yours!" He was holding up a sleek paper plane that looked like it was going to take off on its own.

Erika made a buzzing sound with her lips, holding out a smaller but more thought-out model, and Billy said, holding his wider, larger version of a paper airplane, "Those are dorky; just wait till you see what mine will do."

3

The lives of Tom, Erika, and Billy had been enriched due to their selection by Merlin and Dr. Janus, time travelers, to explore the human intestinal tract a few months back. The trio was still in awe over the adventure they experienced as the "Three Musketeers" of Willowdale High School.

"Man, Dr. Janus' cough is getting worse," Tom said, flexing his growing muscles.

"I'm worried," Erika said. "As smart as he is, being a real time traveler, you'd think he would know to go to the doctor and get treated."

"No worries here," Billy said, waving his hand in an unconcerned way. "I'm not worried, he probably just needs rest. Now let's get to Dr. Janus' assignment," he said, holding up a slick-looking paper airplane.

Erika, a high-spirited red-headed wiz with a sharp wit, held up her plane. It was designed to go high and descend in a long arcing path. She loved hanging out with her buddies, Tom and Billy, but she was also a member of the school's volleyball team and sang in the school's choir.

Tom, the group's football player, held his plane up to the light and openly admired that it looked like it would take off on its own. Recently he had been preparing for the one-act thespian contest at regionals and constantly rehearsed his lines with his friends in a comedy he was a part of.

Billy then showed his plane to his buddies. It was built with a weighted nose so that it would fly faster and farther. He was an active member of all the school's honor societies and loved video games when not working on schoolwork or getting into mischief with his two best buddies.

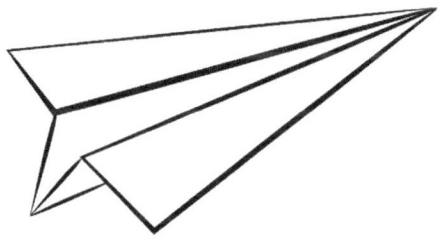

Erika held up her plane to the sunlight as she responded. "Yeah, you are probably right. He hasn't taken a day off to stay home and take care of his body. Okay, we need to work on our physics homework. Let's test these babies. Throw on the count of 3….2….1…"

They had just launched all three at once when a bright shimmering light appeared around them, freezing all motion, including the airplanes in midair. Dr. Janus and the intriguing Merlin, the wise and mysterious guide who had previously taken them on an incredible journey through the digestive system, appeared out of nowhere.

Tom, Erika, and Billy gasped in surprise.

Merlin had a big smile and laughed. "Sorry to surprise you like that, but I felt it necessary to tell you that your next adventure came sooner than expected. Dr. J," pointing at their high school science teacher, who looked drawn and ill, "has had no help from a handful of medical experts, and due to its seriousness, we think it calls for an 'all-hands-on-deck' call to arms to discern what the trouble is."

Merlin's expression changed to one of worry. "We need your help."

Erika asked incredulously, "What can we do?"

"Okay, you know the routine now after our digestive trip. We want you to get inside J's respiratory system and find out what is wrong and fix it." Merlin looked at the three as Dr. J nodded in agreement. "You've proven yourselves as brave and capable adventurers, and we both agree that you are the perfect candidates to mend his illness.

"We think it could be something alien he brought back from one of his many time-jumping adventures as no medical specialist has been able to diagnose, much less cure it. It is getting worse rapidly, and we need you to start ASAP. Are you ready now?"

The three looked at each other—concerned looks on their faces.

When the realization of a new adventure struck the three of them, Tom stood on his toes in excitement, which extended his six-foot frame to a new elevation. Erika grinned big and nodded, which made her ponytail bob up and down, and Billy, with his high IQ, didn't have to say a word; he simply smiled widely because he was always ready for adventure.

As Merlin talked, Dr. J coughed one hacking cough after another, briefly turning blue as he caught his breath. He coughed green mucus into a tissue and looked as if he was about to pass out.

"This trip into Dr. J's system will be different than before because you will not just be learning; you will be saving his life." Dr. Merlin continued. "I hope you understand that I need to be here with Dr. Janus to help if things get serious. I worry that they may."

An amazingly slick-looking high-tech craft appeared as Merlin continued. "I will act as a living hologram inside your vessel as I guide you through the organ system. This newly created transportation is fully loaded with every tool imaginable to protect you and rid Dr. J of this pestilence within his respiratory system.

"Before you get on board, I want to give you a heads-up on what you will be seeing and passing through." At that, a large hologram appeared of a nose on a face. "I will give you the 'lay of the land,' like a road map here."

Billy said, "We were hoping that we would be traveling through the respiratory system, so we have been reading up on it. We have learned how important it is." He opened his iPad, and a similar scene appeared of a nose on the screen.

"Billy, this is no less than I expected of you guys. Regardless, here is what you will see. These openings in the nose are 'nares,' and the mucus-covered hairs are cilia. You will proceed through the pharynx, through the larynx, into the trachea, and pass the epiglottis, a little door that keeps food out of the trachea. You will travel through the trachea, or windpipe, into a bronchial tube that divides into two. Each carries your breath in and out of the lungs.

"Once you reach the finale of the respiratory system, you will meet and observe the alveoli and see how these tiny sacs of tissue function. It will blow you away!"

They all laughed and stopped quickly when Dr. Janus began another coughing fit.

As soon as Dr. J. stopped coughing, Merlin quickly continued. "So you will not be flying blind. You now know what to expect. Since I will be along with you virtually, I will tell you what to do with the Torpedo's arsenal, if needed. This way, I can be in two places at once."

They turned their attention to their new craft, with its clear bubble top and sleek design, much more advanced than the one they used in the digestive system.

The trio climbed in, taking their places in the seats that formed up around them to keep them from falling out and designed to keep them safe, even if the passengers were turned upside down.

Merlin handed each of the trio two gummy bears. One was to be saved for the end of their adventure in order to return to normal, and the other was to be used to shrink in size. They all quickly raised their shrinking gummies as if making a toast and began to shrink as they chewed as Merlin pushed a button on what looked like a phone that had been freed from his pocket.

Suddenly, with a THWUNK sound, the craft and the three occupants shrank to the size of a microscopic grain of sand. Tom, Erika, and Billy piled into the cabin of their vehicle, ready to start this new adventure.

# CHAPTER TWO

## Into the Nasal Passages

Tom, Erika, and Billy sat comfortably in the improved Torpedo craft with a bubble top providing a clear view of their surroundings. The sleek craft was equipped with an arsenal of weapons, such as medication and vitamin supplements that could be used to boost Dr. J's healing capacity. Basically, it was ready for any contingency.

Merlin's hologram flickered to life by the control panel. "Welcome, adventurers," Merlin said in his deep, soothing voice. "Our mission is to find the cause of Dr. Janus' worsening cough and do everything in our power to rid him of it. Be vigilant and ready for anything.

"I know you are concerned after the bumpy ride you experienced on your last voyage inside him, how you may be thrown about due to a sneeze or cough. Do not worry, as I have placed him into a Brachilator, a medical device from the future that will contain any urge to cough while you are inside his body. I would never risk your health and lives if there was not such a device.

## Respiratory system

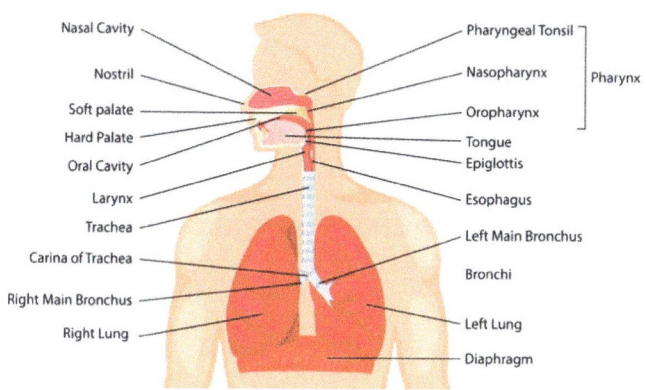

The image you see in the hologram shows the area you will be traveling through. It is important to know what is in the nasal passage that you will have to pass through to get to the lungs. There will be another voyage you will make in the future to learn more about the details of the ear, nose and tongue, so we won't cover the anatomy of these structures. All you need to know at this time is that you will pass through the throat to get down into the trachea, also known as the windpipe. As you can see, the breath enters the nose or mouth, travels through the nasal passages, and into the area where food is separated from the air going to the lungs. It is quite a balancing act to make sure food doesn't get into the windpipe. Somehow mother nature designed the two organs to coincide so closely, yet safely."

Tom just shook his head. "How in the world could all of that take place at once? Eating and breathing at one time. Imagine that three of us, the Musketeers, having pizza, drinking sodas, talking, and breathing all at the same time. And nobody chokes on the pizza! We do this without thinking about what we are doing."

"You are correct, Tom, and as you guys say, "You ain't seen nothin' yet." The expression sounded out of place coming from Merlin's deep, sophisticated voice.

## What Happens to the Breath

### ALVEOLUS GAS EXCHANGE

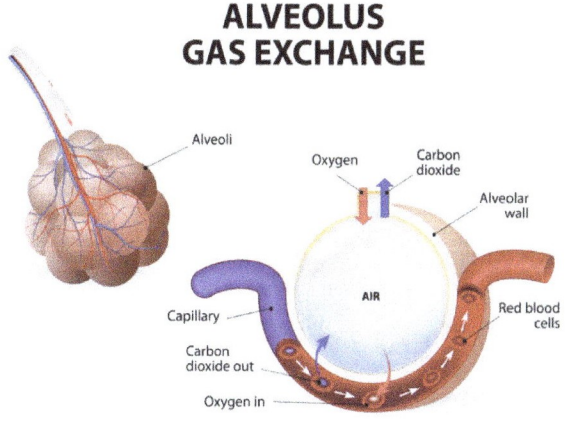

"Before we launch, here is what breathing does and the vital role it plays in sustaining life. Breathing is the first thing you do when you wake in this life and the last thing you do when you leave it. Every day, you breathe in and out about 5,000 gallons (18,925 liters) of air. This supplies the body with oxygen, which is needed to break down food and release energy in the cell. Second, it expels carbon dioxide, a waste product of cellular activities such as making energy.

"The air we breathe each day is made up of a number of different gases, such as nitrogen, oxygen, and hydrogen. Oxygen is about twenty-one percent of the fresh air that is drawn into the lungs when you inhale. After inhalation, it passes through a boundary in the lungs and then enters the blood through tiny arteries called capillaries. These capillaries are the smallest type of blood vessel and surround each alveolus found in the lungs."

Billy said, "We have been studying the heck out of this stuff and know many answers, but just for the record, since everything being said and viewed is recorded, tell us how the lungs are protected."

Merlin said, "Good question. This will enhance the road map you will be traveling."

The hologram grew to become a huge image of a pair of fully functioning lungs located within the protective rib cage made up of twelve pairs of bones, spine, and the breastbone, also known as the sternum. Blood vessels and muscles between the ribs were also shown.

Merlin said, "As you can see, the lungs are well protected just by being within the cage of ribs. The ribs themselves are held together by a strong and flexible set of what is called intercostal muscles. There is a set on the inside behind the ribs called the intercostal muscles and another set on the outside or external intercoastal muscles. These muscles expand and contract with each breath and provide major structural strength to the ribcage. The lungs are well protected within this powerful structure of bone, cartilage, and muscle.

Merlin explained. "There are two lungs, with the one on the right consisting of three lobes, one on top of the other. The left lung has only two lobes. The heart is firmly wedged between the two lungs and is primarily placed on the left side, thus there only being two lobes on the left."

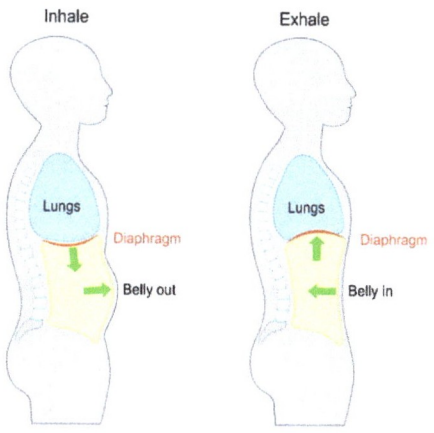

He continued. "Also, between the lungs lay the trachea, esophagus, and a multitude of large and small blood vessels and nerves. Along the bottom of the lungs is a dome-shaped diaphragm. It forms the floor of the chest cavity and the roof of the abdominal cavity and is a type of involuntary muscle that both protects and plays a role in respiration. It has a few openings, one where the esophagus connects to the stomach and another where the blood vessels pass through.

"The blood vessels between the lungs include two large pulmonary arteries that move carbon dioxide-loaded blood to the lungs and two pulmonary veins that carry oxygenated blood from the lungs to the heart. Any more questions?"

Seeing there were no more questions, Merlin said, "OK guys," his hologram pulsing with energy. "Let's Do It!"

The trio fearfully ducked as a giant hand descended down upon the spoon in which the craft had appeared in, now occupied by our adventurers. The craft was carefully placed into Dr. Janus' right nasal cavity, and he was instructed to breathe in easily. They suddenly flowed up into the nostril in Dr. Janus's indrawn breath.

They were surrounded by a dense forest of mucus covered nasal hair, each as huge as a great tree inside the nose. Merlin began to explain about the nasal septum.

"The nasal cavity is divided by a wall called the nasal septum. It is made up of bone close to the skull and cartilage farther away from the skull. If you touch the end of your nose, notice that you can wiggle the end of your nose around. That's the cartilage that usually breaks when someone falls and breaks their nose."

Tom grimaced and said, "Do you guys remember when I broke my nose playing football without a helmet? The doctor had to reset the cartilage, and it really hurt."

Erika looked at Tom with empathy and followed up. "I sure hope you remember to wear a helmet in the future. You had black eyes and a swollen face for weeks after that!"

Tom responded resolutely, "Yeah, I always wear my helmet now."

As Erika and Tom spoke, Billy gingerly touched the end of his nose and moved the cartilage around. Each laughed and giggled when Billy started to make honking noises as he pushed on the tip of his nose.

As the three passengers continued their journey, Merlin described what they were seeing outside the impenetrable but clear hull. "The nasal cavity is lined with tiny hairs called cilia," Merlin explained. "These hairs, along with mucus, function to trap dust, pathogens, and other particles you don't want in the lungs. It usually does a wonderful job keeping those items at bay, but periodically, substances manage to pass through."

The cilia moved rhythmically with Janus's breathing, creating a wave-like motion that reminded Erika of seaweed flowing in the ocean. The walls of the nasal passages were covered in a thick layer of mucus, glistening in the dim light from the outside, growing darker inside the nasal area as they proceeded on their way. The mucus and trapped particles traveled toward the back of the nasal cavity at the top of the throat to a structure called the pharynx.

"Look at the way the cilia move," Billy said, his analytical mind absorbing every detail. "They're like tiny conveyor belts, laying down one way with inhalation and laying the other way with exhalation."

As they moved deeper, they encountered a sudden rush of air as they heard crackling and wheezing. "That's Dr. Janus breathing heavily," Tom said. "It sounds like it's getting harder for him to breathe."

Tom gripped the controls to steady the craft. The air was warm and humid, a testament to the nasal passages' role in preparing the air for the lungs. "I hope he can control his cough and not sneeze while we are in here; we could get smashed in the convulsions and be blown to bits in the explosive winds."

"I am sure the Bronchilator will control any coughing or sneezing, but that is why I am having the hologram ride with you. I am staying here just in case his illness overrides the machine. I am standing right beside him through your trip into and within his body," Merlin said, speaking through the hologram of himself.

Approaching the back of the nasal cavity, the trio watched in delight as they began to move downward, moved through the pharynx, and saw the epiglottis opening and closing, giving a view of the vocal cords of the larynx.

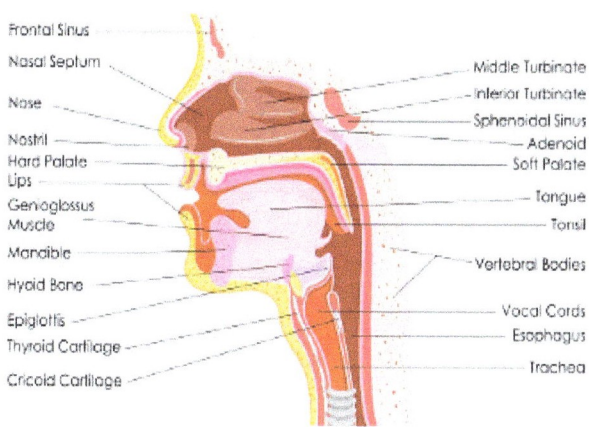

Merlin continued "You saw the importance of the epiglottis, first-hand, in your digestive journey. It closes to allow chewed-up food through and opens when you breathe. Somehow your inner engineer knows precisely when to open or close automatically, without you having to consciously think about it. Now, let me explain how we talk."

Merlin explained. "The larynx is also known as the voice box because it holds the vocal folds, also known as the vocal cords. The muscles and connective tissue that hold the vocal cords in place are protected by a large piece of cartilage called the thyroid cartilage. This is thyroid cartilage is known as the Adam's apple, which is commonly larger in boys."

Tom and Billy instinctively rubbed their throats where the Adam's apple lies as Billy requested, "Merlin, can you get Dr. J to hum lightly so we can see his vocal cords move?"

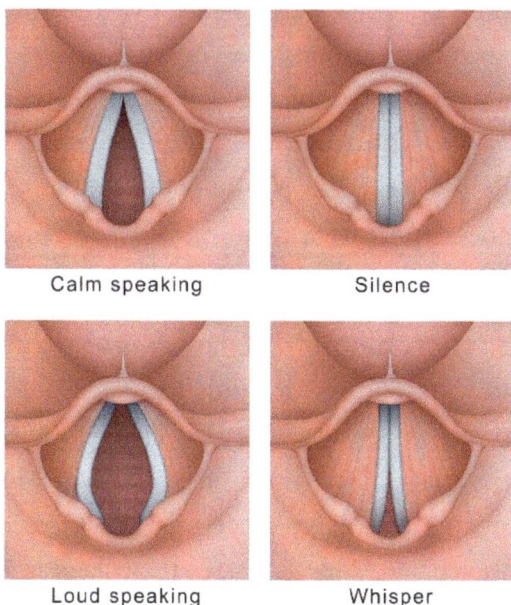

Calm speaking    Silence

Loud speaking    Whisper

Merlin asked, "Are you sure? It will be quite loud."

All three nodded their heads, and soon the vocal cords were moving back and forth. Although Dr. J was humming quietly, it sounded like eruptive booms, so loud the trio put their hands over their ears.

Tom yelled, "Dr. J, please stop! It is too loud!"

After the noise quieted down, they all sighed loudly in relief, and Merlin said, "The vocal cords are relaxed during normal breathing, but when air is forced through the area between the folds, noise is created."

As Merlin finished, Erika began singing the popular song, "Perfect," and said, "I wonder what my vocal cords look like while I am singing in the choir. Oh my, I just had a mental picture of all those vocal cords singing without our bodies. I know that's creepy, but I find it hilarious!"

Tom and Billy looked at Erika as they laughed at the thought of a bunch of bodyless vocal cords singing together.

"Peeeyew," Erika said, frowning, looking at Billy. "Was that you?"

"Not me," Billy said, looking at Tom, who seemed to blush as he looked away.

Then Tom said, "I'm sorry, I shouldn't have eaten that big bowl of beans, I didn't know we would be in such close quarters. I love beans."

"Hey, Merlin," Billy asked the hologram. "I guess you overheard that exchange, and although we are annoyed, Tom is forgiven. There was no place to go to let out that fragrant flatulence." Billy continued. "I am sure we will all be guilty of that sooner or later."

"You are going to have questions from now on and I will answer to the best of my ability as you are in pursuit of knowledge about the human body. It functions as a machine, mostly automatic. It almost seems that there is an engineer in charge of all functions, like respiration and digestion. So, Tom's exhalation from the other end can be excused. Never blame another when such happens as it is entirely natural."

They all laughed at that, and Billy whispered, "Is there a restroom on board?" Then they laughed some more.

Billy frowned and said, "Peeyou," now that we are talking about smelling, can you show us while we are how now we smell stuff?"

"Perfect question, and timely," Merlin said.

## OLFACTORY NERVES

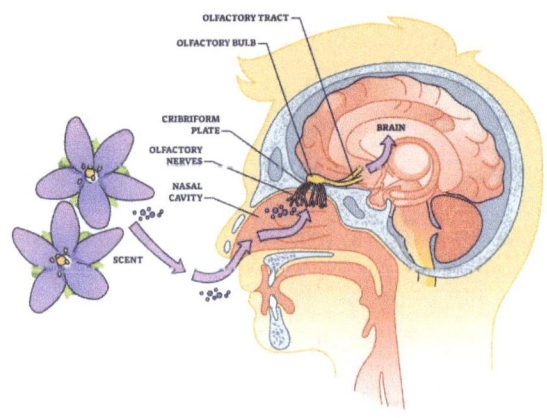

"Here is an image of the olfactory nerves in the nasal area," Merlin explained, "and you probably never really thought about why you have a sense of smelling odors, fragrances, good and bad."

Tom answered as he rubbed his chin. "I can think of a couple. It is a survival mechanism to detect danger, smoke, and predators. Oh, and whether food is fresh or spoiled. Smell also helps you enjoy food, coffee, flowers, and decide if you smell clean. Oh, and it can also trigger memories of things that happened."

Merlin said, "There are many uses of smell in animals, such as reproduction in identifying potential mates, signaling fertility in some species, and in maternal behavior, allowing mothers to recognize their offspring. Animals use smell much more than humans. Perhaps we once had such a heavy reliance on smell, and it was a more reliable resource at some point. All senses are all related to survival."

"Next stop, the trachea," Billy said, his eyes shining with excitement.

The craft was pulled into the trachea, past the flapping epiglottis. Merlin's hologram provided a detailed overview of the structures. Merlin commented, "The epiglottis acts as a gatekeeper; it prevents food and liquids from entering the respiratory tract. During your last trip, you saw this from the digestive point of view. The food you were following traveled down the esophagus and once again you will travel past the epiglottis, but this time, into the trachea as you pass deeper into the respiratory tract.

"The trachea, or the windpipe, transports air to and from the lungs. As you know, the esophagus carries food to the stomach and now you are experiencing the trachea, the hollow tube that carries air to the lungs. One carries food and the other your breath. It is a rigid tube, reinforced with cartilage rings," Merlin explained. "It ensures the airway always remains open. It runs parallel but in front of the flexible esophagus."

Tom exclaimed, "I see rings, rings of cartilage! The rings keep it open so we don't have to think about keeping it open while we breathe."

Billy exclaimed, "Fascinating!"

Erika asked, "Merlin, how many breaths do we usually take per minute?"

Billy proudly answered, "Oh, I know! It is between twelve and eighteen times per minute, but it varies based on your age. Babies and little kids breathe between eighteen and thirty breaths per minute."

Merlin excitedly responded, "That's great work. It is clear that you all have been spending time learning about the human body!"

They continued down the trachea, admiring the layers of cilia and mucus that protected the structure and lungs, although the mucus layers seemed to be getting deeper.

Eventually, the trachea branched into a pair of tubes, one going left and the other to the right, each division leading to the lungs. Merlin told Tom to steer toward the right and into Janus's right primary bronchi.

Merlin followed up. "When the trachea splits, it forms bronchi. The first and largest segment is called the primary bronchi."

"We are now entering the right primary bronchus," Tom said, steering the craft with precision.

They followed the bronchial tubes as they branched into smaller passages, leading the trio deeper and deeper into the lungs. Eventually, they came to a vast area that had passageways that led to structures that looked like clusters of grapes or groups of balloons.

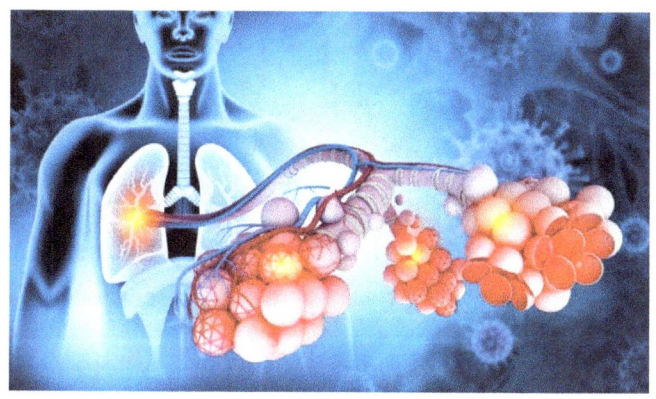

Merlin described further. "These are alveoli. Approximately three hundred million are packed into each lung. If the alveoli from one lung were opened and spread flat, they would cover half a tennis court. With an average adult, the alveoli expand and contract approximately 15,000 times a day. The alveoli remain open until they absorb oxygen, then contract to release carbon dioxide."

Then Merlin asked, "Now, guys, let's think deeply about the two organ systems you have spent time in. How is the function of the alveoli similar to the villi of the small intestine?"

### EXCHANGE OF GASES AT ALVEOLI

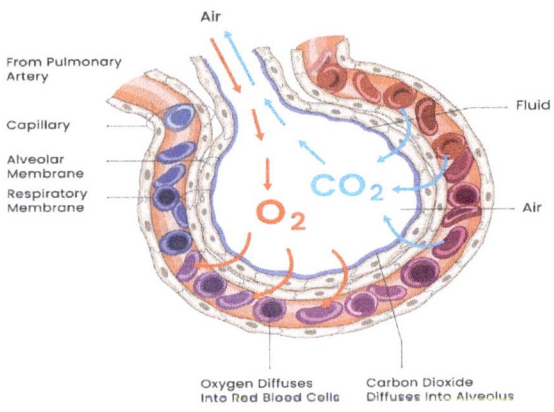

Billy thought for a moment, rubbing his chin. "In essence, their function is similar to the villi and microvilli in the small intestine. Alveoli absorb precious life-giving oxygen into the bloodstream and dump carbon dioxide during exhalation. In the small intestine, villi and microvilli absorb mechanically and chemically digested food and transport it to the blood, then the waste travels to the large intestine to be eliminated through the anus."

Merlin replied with excitement. "I'm impressed, as usual, with you, Billy!"

Then he continued. "The alveoli are where the real action takes place in the respiratory tract. Oxygen crosses through the alveoli's thin, single-celled boundary into the blood to attach to proteins called hemoglobin. At the same time, carbon dioxide leaves the body's blood by crossing that same boundary, traveling through the ever-widening passageways, to the primary bronchi, up the trachea, and out of the nose or mouth."

The further they went, the more mucus they encountered. The craft's sensors detected increased activity.

"There's a lot of mucus here," Billy noted. "It could be a sign of infection."

As they navigated deeper into the lungs, alarms suddenly blared.

"We've got company," Tom said, scanning the monitors.

A swarm of menacing alien bacteria appeared, their strange forms pulsing with an eerie glow. Each microscopic organism was covered in cilia that looked like little arms and legs that moved them around. Each bacterium was splitting into more bacteria as the trio watched the scene in horror.

"Prepare for evasive maneuvers!" Erika shouted.

The craft darted and weaved, avoiding the bacteria. The bacteria emitted a toxic substance, corroding everything touched.

"Deploy the antimicrobials," Billy instructed calmly.

Tom activated the weapons, releasing a cloud of antibiotic agents. The bacteria shrieked and were torn apart, but more scary-looking creatures were advancing upon their tiny craft.

"You need to get to a safer location," Merlin advised. "Head deeper into the lungs, into one of the alveoli."

Tom piloted the craft through the ever-smaller tertiary bronchioles.

Erika cried out, "It feels like we are in tiny caves that are becoming narrower and narrower!"

The bacteria were relentless. Just as they thought they were overwhelmed, Merlin activated the craft's defenses, creating a gooey specialized protective barrier that would keep the trio safe for a few minutes.

"Hold on!" Tom shouted.

The craft sped up, narrowly escaping the bacteria as they entered an alveolus, where the air was a little clearer.

Tom, Erika, and Billy didn't realize that they had been holding their breath during the bacterial onslaught, and each of them breathed deeply, catching their breath.

"Whew. That was close," Erika said as she grasped the arms of her chair, her heart pounding. "But we made it."

Tom said, "Barely," and splayed his legs out, relaxing for a moment.

Billy checked the craft's systems. "We've taken some damage but we're still operational."

"Excellent work," Merlin said. "But our mission is far from over. We must find the source of this infection."

With a renewed sense of determination, the trio prepared for the next stage of their journey, ready to face whatever challenges lay ahead in the depths of Dr. Janus' respiratory system.

# CHAPTER THREE

## The Encounter with Mucklock

The Torpedo craft floated gently in the alveoli, the tiny air sacs that facilitate the exchange of oxygen and carbon dioxide. The trio marveled at the single layer of cells, incredibly delicate structures, each alveolus a tiny sphere of life-sustaining activity.

"Look at this," Billy said, pointing to the monitor, which showed in detail the oxygen being absorbed into the blood. "This is where oxygen enters the blood, and carbon dioxide is expelled. It's incredible."

Erika nodded. "Every part of the body needs oxygen to survive, along with nutrients, and here is where it picks up the oxygen. See those red blood cells racing along those blood vessel highways, this is where they load it on like a truck carrying cargo."

Billy was watching the streams of red blood cells scooting along. "See those cells, they are like red coins and each one  has a kind of sink in the middle, I wonder why?"

Merlin spoke up. "Red blood cells are biconcave, that is they have a cave or sink in the middle. If the cell was flat, it couldn't easily fold over. This sink in the middle allows it to fold into half its size to squeeze through the most narrow capillaries, in order to carry its precious cargo to needed areas, like the smallest toe or hair root. Do you see how it picks up the oxygen molecule and moves off with it? Would you believe that in just over sixty seconds it will be right back here, empty-handed, having offloaded its cargo to pick up another?"

"Jeez," Tom said. "Like a bus picking up its passengers and unloading and the loading again on time.  The human body is amazing."

"Wow! It just hit me!" Erika suddenly said, "If those red blood cells have to slow down for some reason, the body doesn't get its needed fuel and oxygen.  My grandfather has problems with low blood pressure. He says it makes him feel tired and drawn. I bet it is because oxygen and fuel, such as nutrients, can't get to important areas of the body, like the brain!

"He usually rests and feels better." Erika said as she sighed.

Tom said, "Oh wow Erika, what causes his low blood pressure?"

"The doctors aren't sure, but have changed his diet to include nutrients that seem to have helped." Erika answered

Billy responded to Erika's distress "I sure hope that helps your grandpa."

Erika nodded in thanks.

"Okay, guys, we have a job to do. Dr. J is looking worse and worse." Merlin whispered. "We will come back after we take care of his infection so that we can study this incredible miracle of life up close, let's keep moving."

Tom saluted Merlin's hologram and gripped the steering wheel tightly.

Suddenly, the craft's sensors picked up a large, pulsating mass moving toward them through the terminal bronchioles. The screen displayed a creature, unlike anything they had ever seen. It was made entirely of mucus and as big as a slimy house. It emitted an eerie, otherworldly glow.

"That must be the source of the infection," Tom said, starting to steer the vehicle toward it.

A booming voice rocked the boat. "Intruders! Get out of my domain, or I will destroy you now." It headed toward them, slopping and sliding toward the balloon-like alveoli with the evilest intention they had ever felt.

"What is that? A snot man?" Erika questioned.

The trio looked at each other with disgust and started to chant, "Snot Man, Snot Man, you will be destroyed, man!

Merlin's facial expression flickered with concern as part of the hologram.

"I am Mucklock, and this is my world, and I kill all intruders that dare to enter my domain!" Great, slimy tentacles reached out for the Torpedo.

"I am going to infect every living thing on this planet and kill all life, as I did on Moldesto when I caught a ride with your Dr. Janus. He will be my first delicious kill on planet Earth!"

"Snot Man!" They continued the chant, enraging the creature even more. They had hoped that by enraging him further, he would make a mistake. They didn't count on him being nearly impervious to their weapons of medication and vitamin supplements.

"Mucklock's goal is to infect Earth and kill all life," Merlin continued. "He's invulnerable to our current weapons. We need to find a weakness. Get out of there while you can, and we will study the situation for a solution."

# Pharynx Anatomy

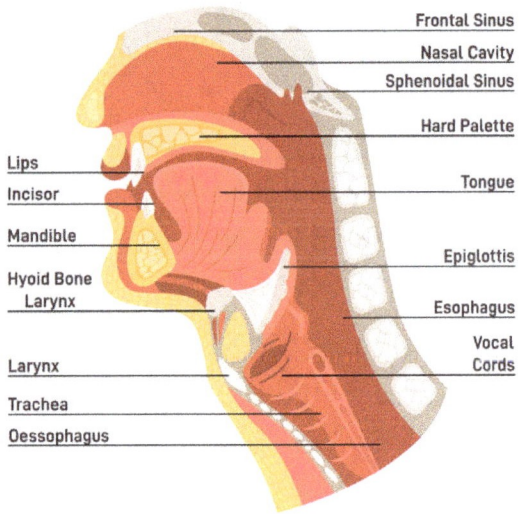

Tom wheeled their craft north to get a better shot at Snot Man as Billy applied nearly every bit of their arsenal toward their adversary.

Suddenly all the tissues surrounding them collapsed in an explosion as they were propelled, topsy-turvy, tumbling over and over up the bronchial tree, through the trachea, larynx, pharynx, and landed upside down in the upper nasal cavity.

Merlin's image popped up, and he said, "Sorry, guys, Dr. Janus coughed and I was lucky to stop his bronchial spasms. All OK there?"

The trio took inventory of themselves, and fortunately, the seats held them tight with a protective covering as soon as they were flipped the first time.

Billy and Erika signaled with a thumbs up they were OK, although Tom had been smacked on the side of his head by Billy's iPad. He groaned as he rubbed his head.

After hitting Tom on the side of the head, the iPad clattered to the floor of their Torpedo craft. When it hit the floor, it opened an app and began playing music which happened to be Billy's favorite song. "Baby, You're Perfect Just the Way You Are" rang through the cabin of the Torpedo.

Erika looked incredulously at Billy, cocked her right eyebrow at him, and reached over to turn off the music.

As the music was turned off, each of the small group looked at each other with concern. A deafening silence filled their vessel as they realized the seriousness of the situation.

After a short pause, Merlin asked, "Did you try everything on Mucklock?"

Erika said, "Yes, we did. The only bit of arsenal that we haven't tried is an anti-parasitic compound, and we are willing to go back and give him a blast of that."

"Well, alright, but be careful and make sure you have room to get out of there if things go badly."

Tom took a deep breath, grabbed the controls of the Torpedo with determination, wheeled them about, and headed back down the pharynx, larynx, and trachea.

Just as they reached the point of the trachea that branched right and left, they spied Snot Man sliding along the tube, picking up mucus and slime as he went, now twice as big as he was when they confronted him in the alveoli. He had grown twice as big as a house, and they all shivered with fear.

"Pathetic humans!" Mucklock screamed. "I was coming to find you, and now you saved me a trip! Get ready to take your last breath." He grew a great slimy tentacle and struck their Torpedo on the side, throwing it sideways. Merlin, operating through the hologram, turned on the anti-parasite gun, coating the slimy creature with blue-green colored slime.

"You pitiful creatures, you are no match for Mucklock, Destroyer of Worlds."

The creature lunged at their craft, spewing more of the thick, toxic, slippery substance. Tom dodged the attack, but the craft was hit, alarms blaring.

"We need to find a way to stop him!" Erika shouted as she frantically checked the systems. "Our weapons aren't effective."

Just then, Billy's iPad, which had been left on the console, accidentally slid onto Billy's lap. Music began to play again, filling the cabin with lyrics. "Be My Girl, I'll Be Your Man" played loudly. Once again, Erika reached over to turn off the music when Billy jerked the iPad away from her with one hand and pointed out the clear roof at Mucklock.

Mucklock, who had been advancing on the craft, suddenly staggered.

"Did you see that?" Billy exclaimed. "The music is affecting him!"

Billy turned up the volume, and the music blared through the speakers. Mucklock writhed in agony, his slimy form beginning to melt.

"It's working!" Erika cheered. "Keep the music going!"

The trio watched in astonishment as Mucklock jerked and spasmed, appearing to have a seizure, and melted away. Mucus spread where he had been sitting and Dr. Janus began heaving as he tried to cough. His throat was loaded up with the mucus that had been Snot Man which was now reduced to a huge lake of liquid nastiness.

Janus began heaving and the Torpedo was tossed around like a pinball in a pinball machine, bouncing from one side of the throat to the other. Soon they were airborne as the ship and a load of mucus was propelled from Janus's lip onto the laboratory floor.

Their tiny ship slid along the lab floor and lay still. Fortunately their seats held them steady although they were bounced around violently. The trio were dizzy but not injured.

Before they could say anything, Merlin scooped them up gently and placed the Torpedo back into Janus's nose and he inhaled them back into his trachea.

"Although you have destroyed Mucklock the Snot Man, the malicious bacteria are still multiplying and could still destroy Dr. J. So get in there and use the Torpedo's antibiotics and scouring tools to clean up the mess. So, get on it and we will talk as soon as your job is done. You are heroes."

The trio cheered with excitement, high-fiving each other. Billy reached over to give the hologram of Merlin a high-five as well. His hand slipped through the image, and he laughed at his mistake.

The immediate danger had passed, but their mission was far from over.

"We need to locate those bacteria that attacked us earlier, medicate the area, provide immune system-enhancing vitamin supplements, and ensure no remnants of the infection remain," Merlin advised.

It took little time and effort to find the offending bacteria. The evil-looking creatures were ten times the size of their craft as they slid on the mucous surface of the respiratory tract.

The bacteria rose up all around the craft, much like a forest of trees reaching from above, and began their attack upon the sleekly created torpedo-shaped craft. Billy deployed the antibiotic that had killed the first group of bacteria, and almost instantly, the bacteria in front of the attack exploded like balloons pierced with needles.

Blast after blast spewed out of the Torpedo and soon there were no more bacteria to contend with. All the pesky beasts had been eradicated.

Tom, Erika, and Billy advanced throughout the respiratory tract, deploying the craft's medical arsenal, and releasing medications into the lungs to decrease mucus and inflammation that had built up due to the infection. They watched in awe as the medications took effect, clearing the mucus and restoring health to Dr. Janus' respiratory system.

As they worked, they took the opportunity to explore the lungs further, observing how the alveoli absorbed oxygen and removed carbon dioxide. With the infection eradicated and Dr. Janus' lungs on the mend, the trio prepared to return to their normal size. It was obvious that his lungs were cleared, for his breathing became easy, with no more crackles and raspy wheeze noises to be heard. They knew their adventures were far from over, but for now, they had saved the day.

"Mission accomplished," the trio shouted, bumping fists gleefully. "The Three Musketeers ride again!"

"Until next time," Erika added, her eyes sparkling with anticipation.

Billy grinned. "We did it together, as always."

## Respiratory system

Respiratory system

Labels (left): Nasal Cavity, Nostril, Soft palate, Hard Palate, Oral Cavity, Larynx, Trachea, Carina of Trachea, Right Main Bronchus, Right Lung

Labels (right): Pharyngeal Tonsil, Nasopharynx, Oropharynx, Pharynx, Tongue, Epiglottis, Esophagus, Left Main Bronchus, Bronchi, Left Lung, Diaphragm

# CHAPTER FOUR

## Merlin Answers Questions

Merlin's hologram flickered proudly. "Well done, adventurers. You've proven once again that there is no challenge too great when you work together. Now, let's explore the area a bit more thoroughly, and I will answer your questions."

"Oh boy," Erika said. "This is the part I love! Here are my questions off the tip of my tongue I couldn't find answers to—"

Tom and Billy interrupted Erika, each raising their hand as if they were in a class asking for attention. "I have some questions," both said at the same time.

Merlin proclaimed, "OK, guys, you have no idea how happy I am at your excitement...OK, Erika, you first."

Erika's great green eyes looked as if they were about to pop with excitement. "Finally, I will get some answers to questions I have always had. What happens when you are 'out of breath,' and what causes yawning, and what is passive smoking?"

"Whoa, whoa!" Merlin stopped her, laughing, holding his hand in a gesture for her to stop and let him answer. "OK, answering your question about being 'out of breath' first. It depends on how quickly the body uses up oxygen and accumulates carbon dioxide. When you are relaxed, you breathe normally. Each breath moves about a pint of air in and out of the lungs. This is about 12 percent of a healthy young adult's maximum breath. In strenuous activity, your breathing may double, and the amount of air taken into your lungs may increase more than five times. You use your lung's reserve capacities by breathing deeper and faster."

"So that's why I breathe hard when running football drills after school," Tom speculated.

Erika agreed. "And I get winded when playing during volleyball tournaments."

Merlin continued. "Yawning is typically not from boredom. It usually occurs after the body's oxygen supply has been depleted after a long period of shallow breathing. This is usually the case when you are tired, under stress, or sitting still for a long time."

Erika excitedly pressed, "What about passive smoking?"

"Passive smoking is inhaling secondhand smoke when you are around people who smoke. Studies indicate many lung cancer cases in nonsmokers are from regularly breathing the air around a spouse or another that smokes."

Tom said, "My turn! What is the difference between a stethoscope and a bronchoscope?"

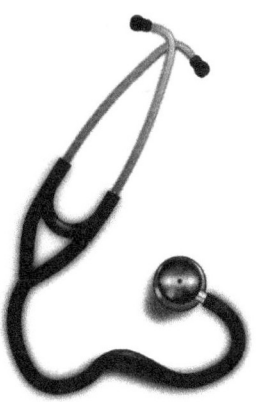

"Some sounds could mean serious disease, so a physician uses a stethoscope as a simple sound conductor. A normal person's breathing has a characteristic sound, and if it is abnormal, they will want to investigate further to determine what could be causing respiratory issues. Abnormal sounds include crackles and wheezes, both of which you heard before destroying Mucklock and Dr. Janus' bacterial infection. Stethoscopes regularly hang around the necks of doctors, so they are easily accessible since they are used as part of the physical exam process to listen to both the heart and the lungs."

Billy commented. "It is super cool to listen to the heart and lungs with a stethoscope! I used one in a lab during biology class one day. I loved hearing the breath sounds of my lab partner."

Merlin continued to inform the trio. "Now, a bronchoscope is very different. It is a long flexible tube inserted into the patient's nose or mouth, which is pressed down into his bronchial tubes to view with a small camera. It allows the doctor to see up close the condition of the organ."

Erika gasped. "Isn't that painful?"

Merlin answered. "Yes, that's why the doctor gives the patient a light anesthetic first because it can cause discomfort. The doctor can then see if there is any irritation or blockage and, in general, make a nonsurgical diagnosis. It has other features like suction, little brushes, and small forceps, which allow a doctor to remove foreign objects in order to fix the patient without surgery."

"Finally, my turn," Billy said. "What causes sneezes and coughs?"

Merlin said, "Interesting, this is a question I hoped you would ask. Pollen and airborne irritants land on sensitive nerve tissues of the nasal area, and a sneeze expels the irritant from the nose. It is a primitive mechanism when the air in the lungs literally explodes, carrying everything that is in its way through the nose and mouth. Each sneeze can contain as many as five thousand droplets unless there is a handkerchief, hand, or arm blocking the explosiveness. Some of this material can carry as far as 12 feet or 3.7 meters."

Tom followed up. "Isn't it true that if an infectious disease is present, it can make someone else ill?"

Merlin replied. "Yes, that's why it is important to block coughs and sneezes from other people."

"Coughing is similar to sneezing as the body is trying to rid itself of irritating material from the bronchi. Physicians pay little attention to coughs until they become regular and may indicate an underlying serious problem such as bronchitis, which is inflammation of the bronchioles, usually due to an infection."

Tom inquired. "I have another question. I have been told that the flu is just a bad cold. Is that true?"

Merlin stopped, looked at the trio sitting and listening to every word, smiled, and continued. "Now, the cold."

He continued. "Physicians define a cold as a viral infection in the upper respiratory tissues and areas that become congested with a stuffy or runny nose. Many times, a sore throat accompanies the stuffiness and congestion. It is a common illness that everyone has sooner or later as an annual affair, sometimes several times a year, particularly in the winter. A runny nose usually doesn't occur until the cold is well underway, sometimes four or more days. It is part of the immune system's defenses. Sneezes are common when someone has a cold because it is the body ridding itself of the mucus and irritants as a result of the cold."

Erika said, "So, is the flu just a bad cold?"

"No, these diseases are caused by different forms of viruses, although the symptoms may seem to be similar," Merlin answered. "The flu may have more serious effects on the body, such as muscle aches, fatigue, and fever as high as 103 degrees Fahrenheit. So it is serious. An annual flu shot is available to those that wish to have it. It is highly recommended that the flu shot be received by people who work in health care, have depressed immune systems, or are very young or elderly."

Merlin followed with, "Do you all have any further questions?"

Billy answered. "What other diseases of the respiratory system should we know about?"

Merlin said, "There are many, such as asthma and emphysema, which are common ailments." He looked away from the trio for a moment, then explained that he would tell Erika, Billy, and Tom all about it as they traveled up the bronchus to the trachea on their way out of Dr. J's lungs.

Merlin explained as the Torpedo traveled slowly up Dr. J's body. "Asthma is a common condition that affects people of all ages. It is a chronic respiratory illness that includes inflammation and narrowing of the airways in the lungs. This can lead to difficulty in breathing and people with asthma may experience symptoms such as wheezing, coughing, shortness of breath, and chest tightness. Sometimes, asthma attacks can be caused by allergens in the air, exercise, cold air, respiratory infections, and stress."

Erika exclaimed, "Doesn't our English teacher have asthma?"

Tom replied, "Yes, she does! So that's why she carries that plastic inhaler around. It must help her to breathe when her bronchi become inflamed."

Merlin continued. "Billy, there is another disease called emphysema that occurs primarily in the lungs of smokers. They tend to have shortness of breath and painful coughing. Emphysema is a serious lung disorder because the alveoli lose their elasticity and blood supply, so the surface area is severely decreased. When the surface area is decreased, there is less oxygen getting into the blood supply and more carbon dioxide stuck in the blood, which leads to an increased chance of infection."

"Oh yeah." Tom threw up his hand. "I thought of something else—what if someone chokes, what do you do? I am not sure I understand how the Heimlich maneuver works."

Merlin stated. "Choking is caused when a foreign body, like food, gets stuck in the windpipe. It could cause asphyxiation and kill you within four minutes. Before it was created by Dr. Heimlich in 1974, choking was ranked as a major cause of death in the United States. Because of that fact, it is important that you know how to perform this maneuver to help someone who cannot talk or cry out because food is blocking the trachea."

Billy asked, "Merlin, can I explain how to do it?"

Merlin replied, "Sure, Billy."

Billy smiled broadly. "First, you stand behind the person, reach around the trunk, and place both hands in a fist right below the rib cage. Give a sharp, hard tug to squeeze the object out. If you are alone and begin to choke, lean over a chair or something with a hard back, and push against it hard and quickly over and over until the object is expelled."

Merlin elaborated. "There are two other things you can do. Put the person's head below his chest and have him lean over, and you pound them on the back between their shoulder blades. Regardless of how you perform the Heimlich, always call 911 for support.

"Another way to save a life is called artificial respiration, or breathing for someone else in need who has stopped breathing. Stretch the victim on their back, lean the head back, pinch the nose to stop air from escaping as you put your mouth around theirs, and blow 12 times a minute for adults and 20 for children. Blow hard enough to raise the chest, like blowing up a balloon. Keep it up as long as you have strength and stamina, for you still may save a life through your endurance."

"OMG," Erika said suddenly. "I thought we had covered everything and then all these new questions appeared in my head. I went snow skiing in Colorado, and for the first day, I was weak and felt almost ill. What was that?"

As their little vehicle reached the back of the pharynx, Merlin answered. "Most people live between sea level and 6,000 feet at the highest all their lives. When you approach heights above 6,000 feet, there is less oxygen available, which may make you feel sick at first. It usually takes a few days for the body to adjust to the decreased levels of oxygen in the atmosphere."

Merlin then questioned, "Anything more, gang?"

They shook their heads as if their current appetite was filled with knowledge, and they were ready to get back into the full-sized world they were accustomed to.

# CHAPTER FIVE

## Summary and the Promise of More Adventure

"If there aren't any further questions left, let's review what you all have seen and learned about the respiratory system. Erika, do you want to start?" Merlin questioned.

Erika gulped and then gushed with excitement as she had a chance to really explain what she had learned that day. "Sure. First, we entered the nasal cavity and watched how cilia and mucus caught potential pathogens." She continued. "We also witnessed the olfactory nerves and understand how smells are detected on the roof of the nasal cavity."

Billy interjected. "We then traveled past the epiglottis into the cartilage-protected trachea after traveling through the pharynx and larynx, where our eardrums were assaulted by Dr. J's humming vocal cords."

Merlin laughed. "I warned you that it could be really loud."

The trio all nodded in agreement as Tom took over. "We traveled to the bronchus and moved through smaller and smaller bronchi until we reached the bronchioles and slipped into the alveoli when we were trying to evade those funky-looking bacteria."

"We then eradicated Mucklock," Billy interjected.

Erika shivered as she chimed in. "He was really scary."

Billy and Tom animatedly nodded their heads as Tom agreed. "Yeah, he was."

"The alveoli are where oxygen and carbon dioxide are exchanged based on the body's needs. After witnessing this amazing and nearly magical view, we traveled back the way we came, and here we are, back in the nasal cavity," Billy explained.

Merlin smiled, laughed, and clapped his hands as he said, "Good, good. You got it! You have seen and now have a deeper understanding of the respiratory system. Let's get you guys back to your airplane throwing."

Merlin said, "Here you go!"

They were jetted out of the nose onto a white Kleenex, chewed upon a different colored gummy bear, and within moments began to grow back to normal size along with the Torpedo.

As they grew back to normal size, each regained their balance and took a deep breath of relief. Dr. Janus was dancing around, and he thanked them profusely as he grabbed each one of them and gave them big, hearty hugs.

They celebrated that evening with pizza and soda at their favorite pizza place. Billy looked inquiringly at Merlin and asked, "What's next?"

Merlin asked, "What do you want to do?"

Erika raised her hand. "Our last two ventures basically converted substances to be used by the blood in the body. Can we travel with a red blood cell in the body and track it? We can observe the arteries, veins, and capillaries, and we can travel with the red blood cells to observe how they work."

"Cardiovascular or circulatory system," Merlin said. "OK, you are on, but I want you to have time to absorb what you just learned during the past two adventures. Not to worry, I will be keeping track of you."

Merlin's and Dr. Janus' shimmering gateway appeared and they disappeared through it as if it was a doorway, and suddenly they were back at their starting point, with their paper airplanes still flying. They looked at each other in wonder at another of Merlin's and Dr. Janus' phenomena, knowing they would show them how to do it when they were ready.

# Glossary

| Abdominal cavity | 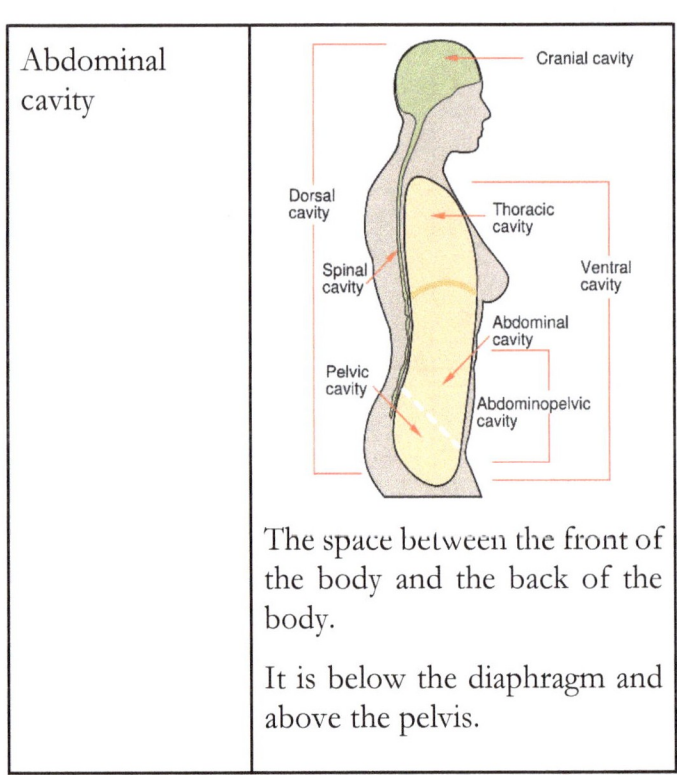 |
| | The space between the front of the body and the back of the body. |
| | It is below the diaphragm and above the pelvis. |

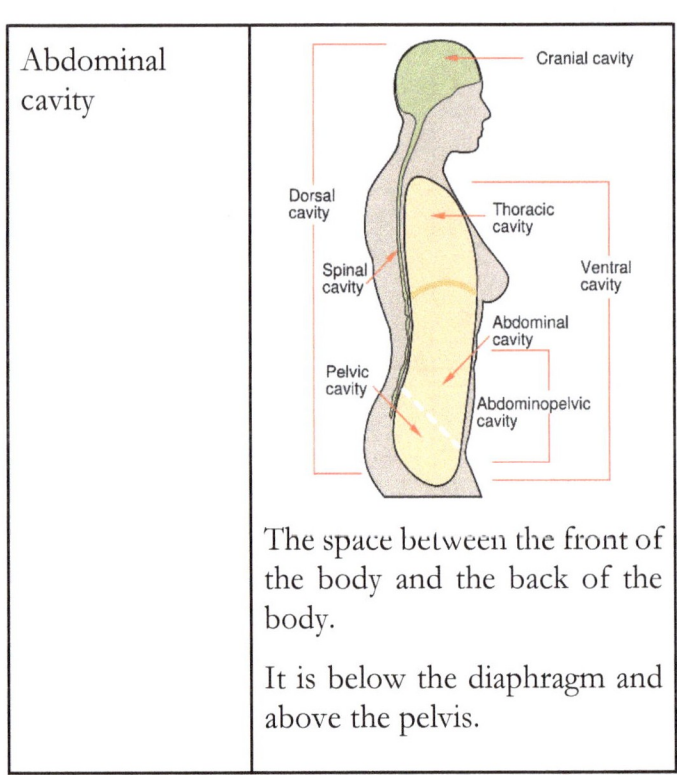

| | |
|---|---|
| Adam's apple | A portion of the thyroid cartilage in the neck which tends to bulge out further in males than in females. The thyroid cartilage protects the larynx and vocal cords. |
| Alveoli | Tiny air sacs in the lungs that allow for gas exchange. |
| Antibacterial | A medication that kills or prevents the spread of bacteria. |
| Antimicrobial | A medication that is used to destroy an infection. |
| Antiparasitic | A medication that is used to destroy a parasite. |

| Artery | |
|---|---|
| | 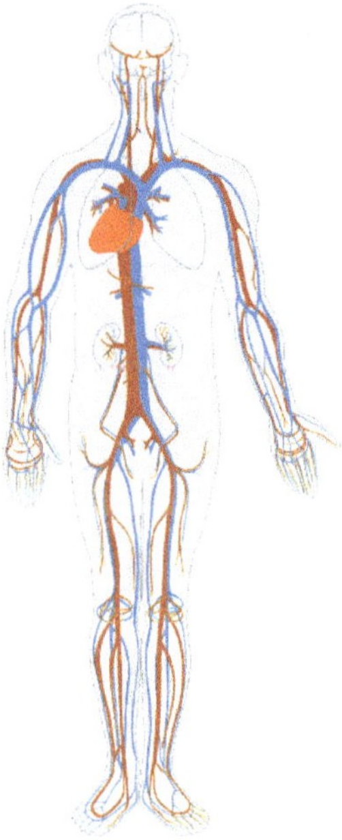 A blood vessel carrying oxygenated blood. The arteries are red on the diagram to the right. |
| ASAP | An acronym that means "As soon as possible." |
| Asphyxiation | Impaired breathing that could cause a loss of consciousness, which could lead to death. |

| | |
|---|---|
| Asthma | A chronic lung disease that causes the airways in the lungs to narrow and swell, making it difficult to breathe. |
| Bacteria | A simple one-celled organism. Most are not harmful and help the human body, but some can cause illness. |
| Breathing | Inhaling and exhaling air to and from the lungs. |
| Bronchial tube/Bronchi/ Bronchus | Airways that carry air to and from the lungs. Become progressively smaller until reaching the alveoli. Primary → Secondary → Tertiary → Bronchiole |
| Bronchilator | An imaginary machine used in this story. |
| Bronchiole | The smallest airway that branches from a larger bronchial tube. |

| | |
|---|---|
| Bronchitis | Inflammation of the bronchi/bronchus. |
| Bronchoscope | A tube with a camera and light that allows a doctor to see within the organs of the respiratory system. |
| Capillaries | The smallest of the blood vessels that carry blood that is oxygenated and remove deoxygenated blood from the tissues of the body. |
| Carbon Dioxide | A gas that is a by-product of metabolism in the body. It is released by the capillaries of the lungs into the alveoli to be released into the atmosphere. |
| Cartilage | A flexible and tough tissue. It is a type of connective tissue found throughout the human body. |

| Cell | The basic structural and functional unit of all forms of life. Bodies are made up of different types of cells. |
|---|---|
| Chronic | A disease that is long-lasting. |
| Cilia | Tiny hair-like particles that can function to move small organisms or substances through the body. |
| Connective tissue | A type of tissue in the body that provides support and protection and binds other tissues and organs together. |
| Contingency | A possible future event that may or may not happen. |

| | |
|---|---|
| Convulsion | A violent movement of the body, usually caused by involuntary muscle contractions. |
| Conveyor belt | A belt that moves items along its length. |
| Corroding | To destroy or damage. |
| Cough | A reaction that causes involuntary muscles to contract due to an irritant stimulating the nervous system to push something out of the lungs or bronchioles. |
| Crackle | Abnormal sounds heard in the lungs during inhalation. Sounds like bubbling or popping. |
| Diagnose | Analyzing a set of symptoms and signs then determining what is causing an illness. |

| Diaphragm | |
| --- | --- |
| | A domelike involuntary muscle that separates the thorax from the abdominal cavity. |
| Emphysema | |
| | Chronic lung disease that destroys the alveoli, making it difficult to breathe. |

| | |
|---|---|
| Epiglottis | A flap of cartilage that protects the respiratory organs from liquids and foods as they pass down the esophagus. |
| Exhalation | Breathing out. |
| Flu | A contagious infection caused by a virus. Also known as influenza. |
| Heart | An organ that pumps blood throughout the body. |
| Heimlich maneuver | Abdominal thrusts that are used to aid a person who is choking. |
| Hemoglobin | A protein found in blood cells that carries oxygen from the lungs to the body. |
| Hologram | An image formed from light. Typically, it is three-dimensional. |

| Hull | The main part of a ship. It includes the bottom, sides, and top. |
| --- | --- |
| Immune System | The body system that protects against foreign substances, cells, and tissues. |
| Inflammation | Redness, swelling, heat, pain, and loss of function due to infection or irritation. It is a normal part of the body's immune response to injury or infection. |
| Intercostal muscles | Muscles between the ribs that play a role in respiration and protection of the lungs. |
| Impenetrable | It is something that is impossible to get through or into. |
| Impervious | Does not allow substances to pass through. |
| Inhalation | Breathing in. |

| | |
|---|---|
| Larynx | <br><br>Voice box that lies between the pharynx and trachea. |
| Medicate | To provide medication. |
| Mucus | A clear fluid that lines many parts of the body. The function is to trap irritants or germs before they invade the body structures. |
| Nares | Nasal passages. |
| Nasal | Pertaining to the nose. |
| Nasal septum | A wall between the nasal passages. |
| Nostrils | Also known as nares. |

| | |
|---|---|
| Olfactory | Relates to the sense of smell. |
| Oxygen | A gas that is required for metabolism in the body. It is absorbed by the alveoli and capillaries of the lungs to be transported throughout the body. |
| Pathogen | A disease-causing organism or agent. |
| Pestilence | A deadly disease that affects a population. |
| Pharynx | **Pharynx Anatomy**<br><br>The area behind the nostrils, mouth, and throat that transports gases, liquids, and foods to either the respiratory system or digestive system. |

| Primary bronchi | The largest bronchiole tube that has split from the trachea. The primary bronchi that carry gas to and from either the right or left lung. |
| --- | --- |
| Pulmonary | Pertaining to the lungs. |
| Pulsating | A strong, regular rhythm. |
| Reserve | To store for later use. |
| Rhythmically |  A periodic movement or beat. |
| Small intestine | A long tube-like organ that is part of the digestive system. It is responsible for the digestion of food and absorption of the digested particles. |

| Seizure | A temporary episode of abnormal activity in the brain. |
| --- | --- |
| Sneeze | An involuntary expulsion of air from the lungs due to an irritation in the respiratory system. |
| Sternum | A bone in front of the rib cage. Also known as the breastbone. |
| Stethoscope | <br><br>A device used to listen to the heart and lungs. |

| | |
|---|---|
| Terminal bronchiole | The smallest branching bronchiole. It carries gases to and from the alveoli. |
| Thyroid cartilage | The cartilage that covers and protects the larynx. |
| Toxic | A harmful substance. |

| Trachea | 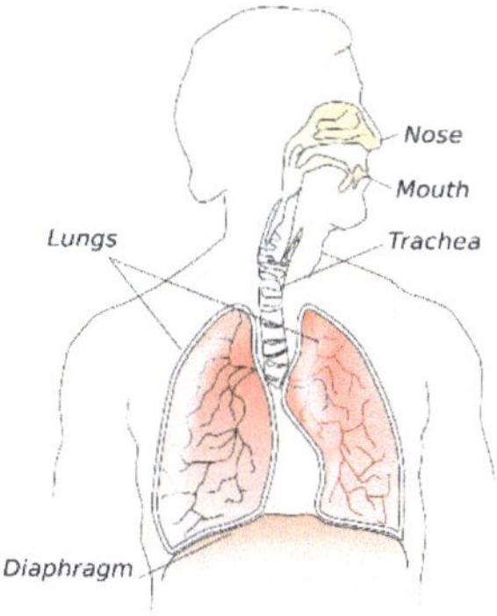 A rigid organ of the respiratory system that carries gases from the mouth and nares to the lungs as well as away from the lungs to the outside world. |
|---------|---------|
| Vein | 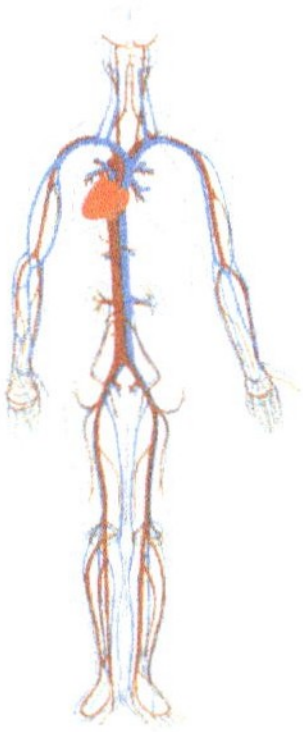 A blood vessel carrying deoxygenated blood. The veins are blue on the diagram to the right. |

| | |
|---|---|
| Villi | Small fingerlike structures on the membranes of organ systems that function to increase surface area and move substances along the surface of the organ. |
| Viral | A disease caused by a virus. |
| Virus | An infectious agent that replicates only inside of living cells. It can cause many diseases. |

| Vocal cord | Calm speaking     Silence <br> Loud speaking     Whisper <br><br> An organ of the respiratory system that allows vibration and creates sounds. |
| --- | --- |
| Wheeze | High-pitched whistling sound due to a condition in the respiratory system. It is usually due to narrowed or blocked airways. |

# About the Authors

L.D. Sledge, an 89-year-old veteran, cancer survivor, musician, and artist, had a remarkable career as a courtroom lawyer for 43 years in New Orleans and Baton Rouge, Louisiana, until 2003. Throughout his career, he tried over 200 judge and jury trials in his own firm. Upon leaving the law, Sledge began pursuing his lifelong dream of dedicating the rest of his life to writing. Remarkably, he had already been writing for over 30 years while practicing law and has now been writing for more than 50 years. With over 3 million words in print, Sledge's work spans every genre and subject, including books, novels, blogs, cookbooks, articles, and even poetry. Reading his work is an opportunity to connect with the words of a man who has lived life to the fullest, offering unique insights and perspectives shaped by his diverse experiences and accomplishments.

Dr. Shannon W. McPherson, a highly trained and experienced high school science teacher, has had twenty-six years of teaching in a wide variety of sciences. These include anatomy and physiology, biology, medical microbiology, chemistry, environmental science, and pathophysiology. She has also been an adjunct faculty member at the college level, teaching anatomy and physiology to premedical students. Dr. McPherson has taught thousands of students at various levels, which attests to her passion, drive, trustworthiness, and love for her students and their needs.

Since retiring from the public school classroom, she has returned to adjunct as a professor at a local college and has opened Motivate Your Mind LLC to guide students, teachers, and fellow coaches in their pursuit of knowledge by motivating and guiding them to reach their greatest potential both in and out of the classroom.